Windows 10

The Ultimate Beginner's Guide!

TABLE OF CONTENTS

INTRODUCTION

I want to thank you and congratulate you for purchasing this book...

"Windows 10: The Ultimate Beginner's Guide"

Welcome to *Windows 10* – the newest operating system in Microsoft's flagship line of products. Released not too long after *Windows 8* and *Windows 8.1,* which garnered mixed reactions from the end users, *Windows 10* cued a 'fresh start' for one of the most important pieces of computing technology today. It delivers the best out of Windows 8/8.1 along with those from *Windows 7.*

Here are some of the topics you will learn from this book:

- Creating user accounts for younger family members

- User account security

- Cortana voice commands

- Personalizing the look and feel of your OS

- Using the built-in recording software to record app runs and gameplay

- And many more!

Thanks again for purchasing this book, I hope you enjoy it!

CHAPTER 1 – WELCOME TO WINDOWS 10!

In This Chapter:

***What's New in Windows 10?*

***Main Differences between Windows 8/8.1 and Windows 10*

***Upgrading to Windows 10 (from Windows 8.1 and Windows 7)*

What's New in Windows 10?

Microsoft definitely added a ton of new features in Windows 10. From a more intuitive user interface design to an intelligent *digital assistant,* you can probably spend an entire week exploring all the new things you can do with Windows 10! But of course, some changes are more significant than others. Here are some of the highlights of Microsoft's new operating system as well as the essential changes between Windows 8 and Windows 10:

- **The Return of the Start Menu** – One of the reasons why Windows 8 and 8.1 were disliked by many is because of the complete elimination of the *Start menu.* And in its place came the *Start screen,* which is apparently more suited for devices with touchscreen displays. In Windows 10, the Start menu makes an elegant comeback.

- **Microsoft Edge** – Previously dubbed as 'Project Spartan', *Microsoft Edge* is Windows 10's brand new internet browser. The icon may eerily resemble the *Internet Explorer* icon, but when it comes to functionality, there is no doubt that Microsoft Edge outperforms one of the most *hated* browsers in the history of computers. It is fast, minimalistic, and responsive.

- **The Settings Menu** – Windows 10 has an all-new *Settings menu* that can give you access to everything you can change in your operating system — all in one place. It is simplistic and segregates all options into categories; namely *System, Devices, Network & Internet, Personalization, Accounts, Time & Language, Ease of Access, Privacy,* and *Update & Security.* It also comes with a search bar to help you find exactly what you're looking for.

- **OneDrive Integration** – Microsoft's *OneDrive* — a cloud storage software — comes preinstalled to Windows 10 systems. Basically, it allows you to save and access anything from anywhere thru synchronization. You may even set your OneDrive storage as the default save location for specific file types, including *documents, music, pictures,* and *videos.*

- **Universal Apps** – Microsoft took a bold step in introducing *universal apps* to Windows 10, which are essentially the same as those being run in tablets, phones, and the *Xbox One* — a home entertainment/games console. Impressively, the *Windows Store* already has countless of downloadable content from Apps, Games, Music, and even Movies.

- **Continuum** – Apparently, computers with touchscreen displays are on the rise, which is why the Windows line of products are moving in a direction that will accommodate computers with this technology as well as those without it. With *Continuum*, Windows 10 users can easily switch between the desktop mode and the *tablet mode*. Basically, it quickly optimizes the system for these modes whenever necessary (e.g. *removing a detachable screen*).

- **Cortana** – Cortana, the acclaimed *Windows 10 digital assistant,* is Microsoft's answer to Apple's *Siri* and *Google Now*. However, a lot of users find Cortana to be faster, smarter, and *more fun* to talk to. There are plenty of things you can do with Cortana from scheduling reminders to joke-telling. More on these on *Chapter 7 – "Hey Cortana"*.

- **Game Streaming** – If you own an Xbox One console, then you can *stream* your game directly into your Windows 10 device; be it a tablet, PC, or laptop. Furthermore, you can download an app that grants access to your messages, friends list, third-party game clients (*Steam*), and everything else in your activity feed. Additionally, Windows 10 supports *DirectX 12,* which can increase your CPU utilization by up to *50%*. Gaming may not matter to you at all, but just bear in mind that Windows 10 is CPU-intensive, which is why this increase in CPU efficiency is much needed.

Keep in mind that most of these changes shall be discussed with further detail in the following chapters. If you already have Windows 10 up and running on your computer, then feel free to skip ahead to the next chapter. But if you still haven't upgraded your system yet, then here are the things you should know.

Upgrading to Windows 10 – What you should know

Perhaps another most important (and surprising) change in Windows 10 is that it comes with a price tag of absolutely free – for the *first year* of its release at least. And you probably know it by now since Microsoft made a particularly big announcement. It is, in fact, one of the biggest software upgrade program in Microsoft's history.

Ever since July 29, 2015, Microsoft began rolling out the free updates to eligible users. If your computer is eligible for a free upgrade, then you should already be aware of the *Get Windows 10* taskbar notification by now. *It is recommended that you upgrade your Windows 10 software through this app.* This is because the Windows Update program will automatically scan, download, and install the updated drivers to your devices for a smoother transition. This is also the reason why you need to have a *stable internet connection* throughout the entire upgrade process, which could last for several hours.

Another issue is that the Windows 10 update may be delayed indefinitely for a lot of people. This is why Microsoft offered an alternative for performing the update, which is done by using the *Windows 10 Media Creation Tool*.

CHAPTER 2 – GETTING STARTED WITH WINDOWS 10

In this Chapter:

***A Look around Windows 10*

***Using and Setting up the Action Center*

***Changing How Action Center Works*

After successful installation of Windows 10, the first thing you need to do is to sign in to your Microsoft account, which will be used for a lot of features such as cloud storage synchronization, email, calendars, and so on. This is *mandatory* for every new installation of Windows 10. If you don't have a Microsoft account yet, then you should create one by clicking "Create a new account" in the setup wizard. Otherwise, visit https://signup.live.com/ to register a new account.

A Look around your New OS

The first thing you probably want to do is to press the Start menu button. You should notice that the Start menu in Windows 10 also has *tiles* from the Start screen in Windows 8, although it no longer occupies the entire screen.

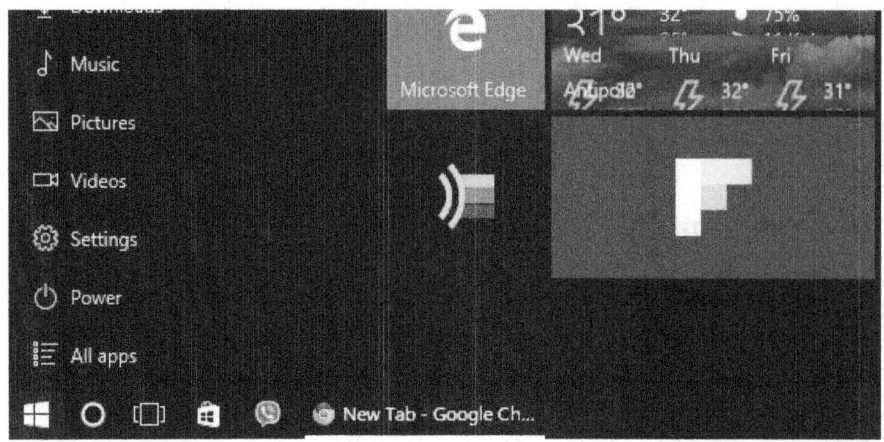

Pressing Settings will bring you to the all-new Settings menu of Windows 10:

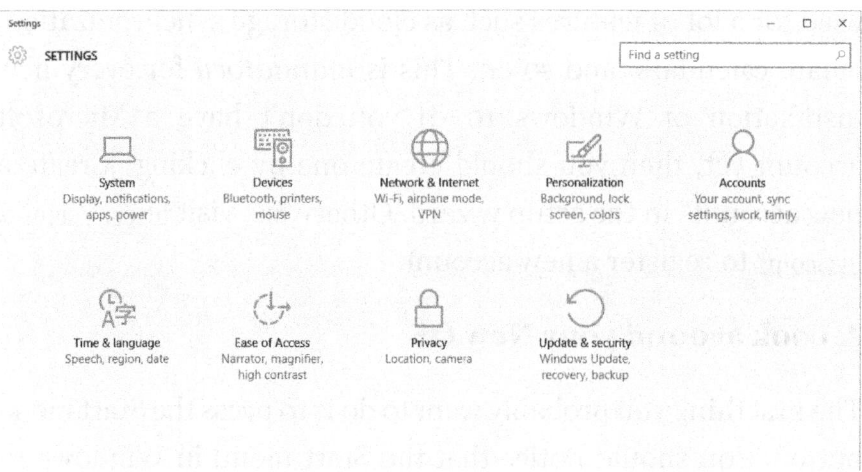

Take note that you may also access this screen through the *Action Center,* which will be discussed shortly. Keep in mind that you will be going back to this screen multiple times throughout this eBook.

A notable difference between Windows 10 and *all* the previous versions of Windows is the use of *plain white title bars* in almost everything. Regardless of the theme or color scheme you use for your computer (see *Chapter 4 – Display and Personalization* Options), the title bars for Windows explorer and other software such as *Microsoft Office applications,* internet browsers, email clients, and the Settings menu (above) will have white title bars. However, you can apply a 'fix' that will make you have colored title bars in Windows 10. This will be discussed in the next chapter.

The Action Center

Another new feature in Windows 10 is the *Action Center,* which is a single location for system-wide notifications, including other accounts linked to your Microsoft account. You can go to the action center by pressing the speech bubble icon in the taskbar. Once pressed, the Action Center should contain all your notifications:

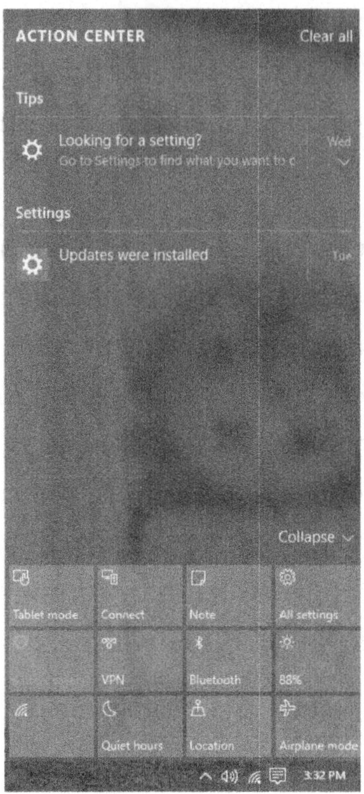

Aside from the notifications, you should also be able to find the following options in your Action Center:

- **Tablet Mode (Default Quick Action)** – As explained in the previous chapter (Continuum), the Tablet mode optimizes the system for touchscreen tablets. You can quickly turn this setting on or off through the Action Center. Take note that the Start menu will use up the whole screen -- just like in Windows 8 -- when in Tablet mode.

- **Connect (Default Quick Action)** – This allows you to wirelessly connect to external devices such as Bluetooth-enabled monitors or speakers. Keep in mind that your Windows 10 device *must* have Bluetooth in order to detect these wireless devices. Additionally, be sure that you have the latest Bluetooth drivers installed and have Bluetooth turned on.

- **Note (Default Quick Action)** – Pressing *Note* in the Action Center will open up *Microsoft OneNote;* a note-taking application for jotting down just about anything you need from shopping lists to personal diaries. Of course, the use of OneNote as your primary note-taking application is optional.

- **All Settings (Default Quick Action)** – As mentioned above, you can access the Settings menu by pressing this tile in the Action Center. There really is no difference between accessing the Settings menu through the Start menu and through the Action Center.

- **Batter Saver** – You can use this option to maximize power efficiency – useful if you want to save your battery. Of course, this option is not available in case there is no battery in your system.

- **VPN** – This will allow you to quickly access a *Virtual Private Network* or *VPN*. For this, you will need to select a VPN provider as well as a valid server name or address and the necessary credentials. If you're trying to connect to a VPN in your office, contact your network administrator.

- **Bluetooth** – This allows you to quickly turn your device's Bluetooth connectivity on or off. Or, you can right lick this tile and choose *Go to settings* to open Bluetooth settings. This is important if you want to manage paired devices or adjust advanced Bluetooth settings.

- **Brightness** – You can quickly adjust the display brightness of your screen using this tile. The level of your brightness should be displayed in %, and pressing this once will increase it by 25%. If it is maxed at 100%, pressing it again will cause your brightness level to readjust to 25%. Of course, you may use other screen settings such as using an *Fn* command in your laptop to adjust your display's brightness manually.

- **Wi-Fi Settings** – You can turn your Wi-Fi connectivity on or off in the Action Center. If you set your device to "Connect automatically" to a Wi-Fi connection in range, then choosing this tile will automatically connect you to it. Otherwise, you'll need to connect first by clicking on the Wi-Fi icon on the notification area of your taskbar (∗).

- **Quiet Hours** – This is a feature that disables all notification alerts such as new emails, app updates, social media updates, and so on. This is just as useful as muting your device.

- **Location** – Turning this feature on allows specific apps like Cortana to use your location. Keep in mind that a lot of Windows 10 apps and features require location verification in order to function correctly. If you see a circle icon (◉) in your taskbar notification area, then an app is currently using your location.

- **Airplane Mode** – Finally, Airplane mode turns off all wireless connectivity to comply with airline safety protocols. This includes Wi-Fi and Bluetooth connectivity.

Changing how Action Center works

Quick Actions (default to Tablet, Connect, Note, and All Settings) can be replaced with any other tile available in the Action Center. Keep in mind that this only affects the topmost tiles in the Action Center. You can do this by going to Start → Settings → System (Display, notifications, apps, power) → Notification & Actions.

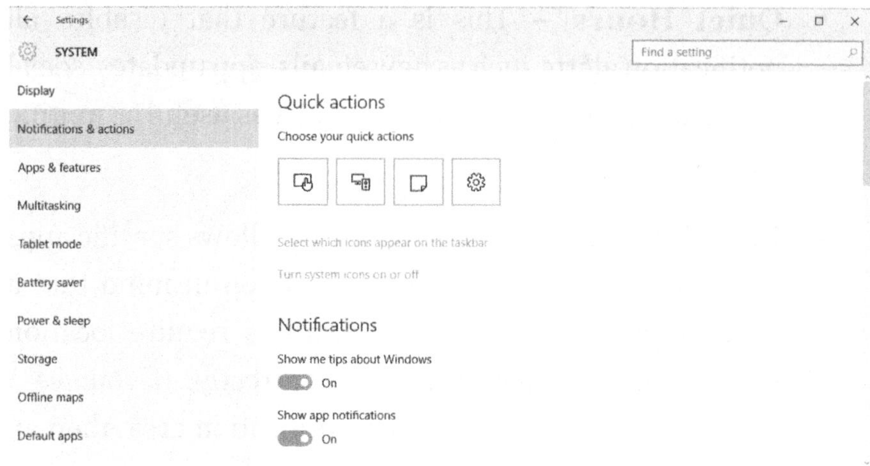

You can control which apps can display notifications in the Action Center here as well. Simply turn notifications off by pressing the 'switch' to the right of an application. For example, say you want to turn off Alarms & Clock notifications on the Action Center:

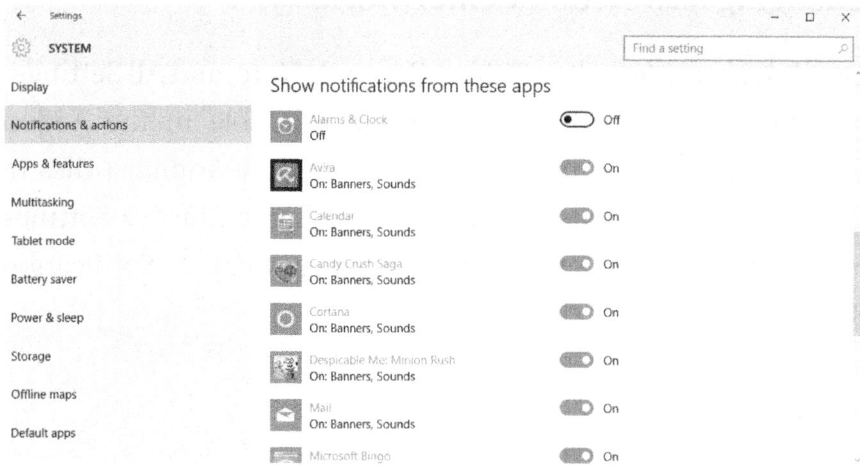

Lastly, you can change the color of the Action Center by adjusting *Personalization* settings, but that will be for the next chapter. Right now, you should focus on getting desktop features such as your Action Center to work to your liking. The next step is to fully manage your accounts for billing info, subscriptions, security settings, and so on.

CHAPTER 3 – GETTING YOUR ACCOUNTS READY

In this Chapter:

**Managing your Microsoft Accounts and Account Settings*

** *Setting up Account Security Settings*

**Creating Accounts for your Family Members*

You can manage your Microsoft accounts by going to Accounts (Your account, sync settings, work, family) in the Settings menu. This is divided into 5 different sections with specific settings each:

Your account

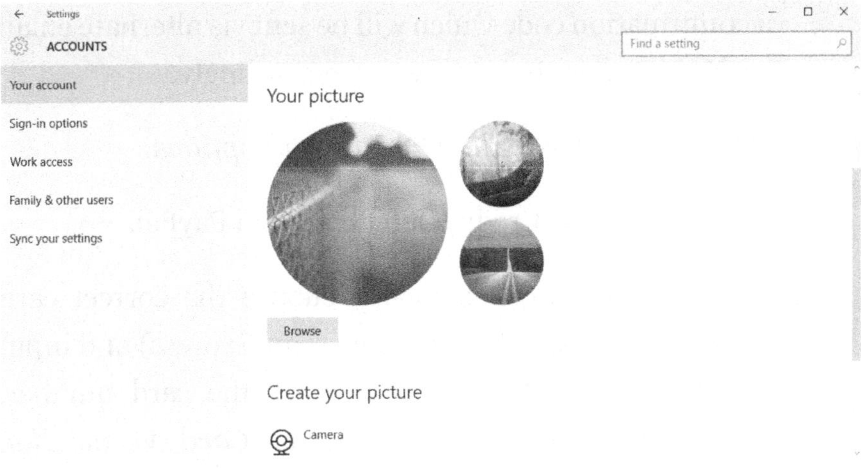

Here, you can set your picture which will be displayed in this device. Additionally, you can manage *additional* Microsoft accounts as well as accounts for work or school. Of course, you can also manage your main account – primarily for the purposes of managing your devices as well as setting your payment options and billing information.

Remember that setting your payment options here will allow them to be accessible whenever making in-app purchases (purchases to be made while inside an application such as a virtual *game store*). This is why it is better to set up your payment options as early as possible. Here are the steps you should follow to setup a payment option:

1. Go to the Accounts settings → Your account page and click "Manage my Microsoft account".

2. Login to your Microsoft account. You may need to input a confirmation code which will be sent via alternate email or SMS. Be sure to set these up accordingly.

3. Go to *Payment & billing → Payment options.*

4. Choose between Credit/Debit card and PayPal.

5. If using a Credit/Debit card, choose the correct card type (Visa, MasterCard, or American Express) and input the necessary information such as the card number, expiration date, name, and CVV (*Card Verification Value*).

6. Supply your complete billing address information including a working phone number and valid email address.

7. Click 'Next' and you will be taken to the necessary verification page of your chosen payment option. Once linked to your account, you can easily use these payment options from within Windows apps and from the *Windows Store*.

Finally, you may also choose to sign-in to your computer using a *local* account and use a conventional *password* instead of a PIN. Simply click on "Sign in with a local account instead" so you can sign-in to your Windows 10 device without having to sign-in your Microsoft account. Instead, you will be asked to insert your current account's local password:

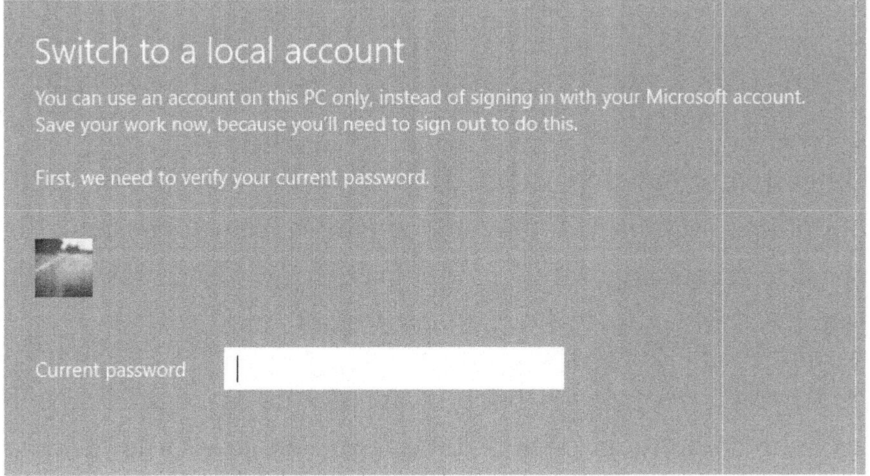

Sign-in options

The next section in Accounts settings is your *Sign-in options*. Under 'Require sign-in', you can adjust when your password or PIN will be requested. This can be set to "When PC wakes up from sleep" or "Never". Keep in mind that your computer will always require a PIN or password whenever you login to your account from startup.

Creating a Picture Password

Other than changing your PIN or password from the Sign-in options section, you may also create a *picture password* in order to create an entirely unique password. All you need to do is to press the 'Add' button under 'Picture Password'.

First, you'll need to verify your account using your Microsoft account credentials. When done, you can press the 'Choose Picture' and then find any photo you'd like to use for your photo password. Bear in mind that high-resolution photos are recommended. Once selected, drag the photo into position using your mouse or the touchscreen display.

When done, you can use a combination of up to *three* gestures – circles, straight lines, and taps/clicks. Remember their exact location, direction, and the order in which they are included in your picture password. Of course, you can start over for as many times as you need. Simply press the 'Start over' button next to 'Cancel'.

Work Access

Here, you can set up access to apps, networks, emails, and other shared resources at work or school. Take note of the existing policies set by your network administrator in these environments regarding your use of such resources.

Family & other users

Your family

Add your family so everybody gets their own sign-in and desktop. You can help kids stay safe with appropriate websites, time limits, apps, and games.

 Add a family member

Learn more

Here, you can easily set up multiple user accounts for different family members. This is a great way to control each member's usage by setting specific time and access limitation. This time, two account types are being used; 'child' and 'adult'. Select the circle next to the account type you're trying to create.

Add a child or an adult?

Enter the email address of the person you want to add. If they use Windows, Office, Outlook.com, OneDrive, Skype, or Xbox, enter the email address they use to sign in.

○ Add a child

○ Add an adult

First of all, keep in mind that the family member you're trying to add must have their own email address (Microsoft account). You can easily create a new account by clicking "The person I want to add doesn't have an email address". This will take you to a new Microsoft account creation page. Here, you'll need to supply the necessary information for a new account (First name, Last name, new email address, Password, and country of residence).

The next step is to add the security information of your family member. You can either use a phone number (default) or an alternate email. This will be used for recovery purposes should that family member loses access to his/her account. Next, you'll have to select the following account preferences you wish to use for your account. Simply select the checkbox to the left and press next to continue:

See what's most relevant to them

Make sure they see the search results, advertising, and things they'll like most when Microsoft personalizes their experiences by using their preferences and learning from their data. Change these settings online and in some Microsoft products and services.

☑ Enhance their online experiences by letting Microsoft Advertising use their account information. (They can change this setting at any time.)

☐ Send them promotional offers from Microsoft. (They can unsubscribe at any time.)

Once you're happy with your account preferences, press next and you will arrive at a verification screen. A child account will automatically have the default restrictions and have their activities monitored through the computer administrator's online Microsoft account. Adult accounts will also be able to *block* child accounts from specific websites, apps, and games as well as set allowed times for usage.

To manage the child accounts in your computer, go to https:// account.microsoft.com/family.

Adding Someone Else?

You may also add another user who is not a member of your family. Simply click on "Add someone else to this PC" below the settings for your family.

Sync your settings

The last section in Accounts settings allows you to synchronize Windows settings across all your Windows 10 devices. However, a device must be logged in under the same account and has syncing turned on before settings can be synchronized. You may also specify the settings you wish to synchronize such as your *theme settings, browser settings, passwords, language, ease of access,* and *other windows settings.*

CHAPTER 4 – DISPLAY AND PERSONALIZATION OPTIONS

In this Chapter:

**Changing and Setting Theme Preferences*

**Personalizing your Lock Screen*

**Enabling Colors for your Title Bars*

One of the biggest changes Microsoft has implemented past Windows 7 is the use of *Metro* – which is basically *flat design*. It was introduced with the release of *Windows 8* and is refined in Windows 10. In Metro, the shaded colors and curvy window borders are replaced by single-tone colors, sharp edges, and a minimalist overall layout. While some people are still adjusting to the new 'feel' of Microsoft's latest operating systems, the user interface of Windows 10 is indeed a lot more intuitive and fast, especially with the introduction of the Action Center, *Cortana,* and the new Start menu.

A Closer Look at the Start Menu

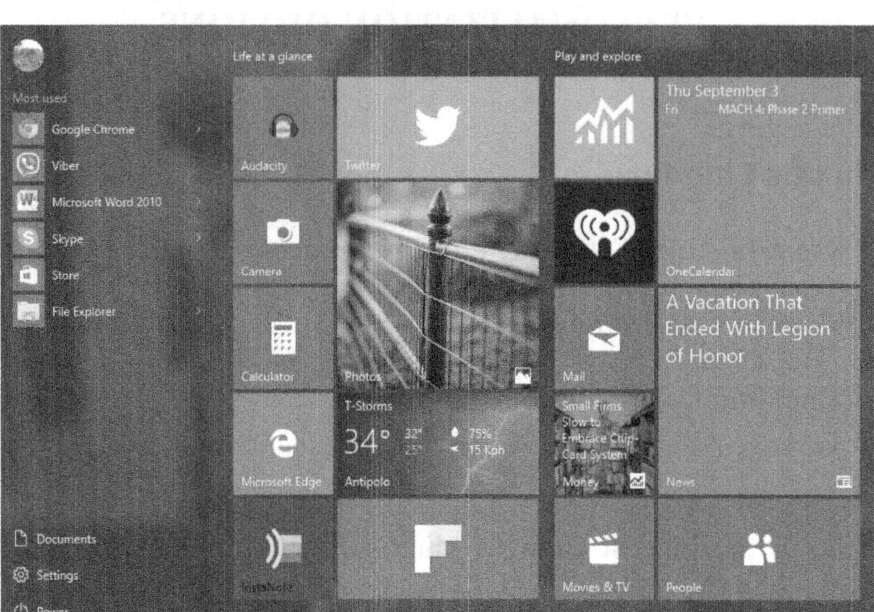

As you can see, the new Start menu in Windows 10 incorporates the tile elements in Windows 8 with the classic menu to the left, which is present in older operating systems like Windows XP and Windows 7.

On the left side, you can browse all the apps installed in your device by clicking 'All apps'. These apps will be arranged alphabetically by default. However, you can search for apps starting with a specific letter, number, or symbol by clicking on top of the Start menu.

You may also specific which folders will appear in the Start menu by going to Start → Settings → Personalization → Start and then click on "Choose which folders appear on Start". Aside from this, you can change the way Start behaves. You can make Start show the most used apps, recent apps, suggestions, and recently opened items. You may also use Start in full screen by selecting "Use Start full screen" regardless if you're using a touchscreen-enabled device or not.

Pinning Shortcuts

You can right-click on any shortcut and choose *Pin to Start* to put them into the Start screen to the right. You can also drag and drop the any shortcuts in order for them to appear in tile view. Tiles can be placed and rearranged freely in any of the tile groups. There are two default groups that contain appropriate apps on the Start menu -- "Life at a glance" and "Play and explore" groups.

- **Life at a glance** – Here, you'll find apps that are related to your social and personal life such as Twitter, *People*, photos, and so on.

- **Play and explore** – On the other hand, apps included in this group are all about exploring the collection of apps that are downloadable from the Windows Store.

Keep in mind that you can rename any group by left-clicking on its title text. You may also add a new group at any time by pinning a new app. You can leave this group nameless, or name it yourself by left-clicking on top of that group. Aside from the Start menu, you can also pin items to your taskbar, which allows quicker access – straight from your desktop.

Tiles can be resized into *small, medium, wide,* and *large.* This allows for better customization on how you want the Start menu to appear. Some tiles are limited to only small to medium sizes, especially third-party apps that did not come from the Windows Store.

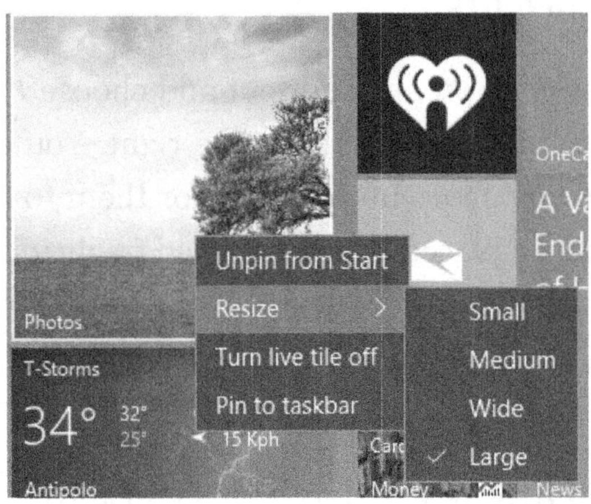

Some tiles also have 'live tile' modes, which will preview notifications or other features of that app. For example, the Photos app will show your pictures in a slideshow when live tile is enabled. You can turn live tile mode on or off by right clicking on the tile you want to change.

Personalizing your Windows

Just like in previous versions of Windows, Windows 10 allows you to quickly select or create a *theme,* which will dictate how your operating system looks. You can access all these personalization settings by going to Start → Settings → Personalization or by right-clicking anywhere in the desktop and clicking 'Personalize'.

Background

First, you can change your desktop background settings by choosing between a slideshow, a picture, or a solid color. Simply select the one you prefer in the dropdown list under 'Background'.

Colors

An interesting feature in Windows 10 is the interface's ability to change color by picking an *accent color* from your current background. This is useful if you have a slideshow for your background. To do this, simply turn on "Automatically pick an accent color from my background".

You may also disable or enable colors for the Start menu, taskbar, and Action Center by selecting or deselecting "Show color on Start, taskbar, and action center". Doing so will make accent colors only affect tiles while Start, the taskbar, and Action Center will remain black. Finally, you can enable or disable transparency for these interface elements by selecting or deselecting "Make Start, taskbar, and action center transparent".

High Contrast Settings

Clicking on "High contrast settings" will allow you to choose a high contrast theme, which is meant to improve the visibility of certain interface elements. True, it may not be the most visually appealing decision to choose a high contrast theme, but it can be useful if you want to reduce eye strain.

Lock Screen

You can personalize your lock screen by specifying a background picture, a slideshow, or by choosing *Windows spotlight.* Choosing this will allow your lock screen background to change each time – adapting whether or not you *like* a specific type of image like nature, abstract, or digital art. This is a good idea to keep your lock screen experience new each time you sign-in to your account.

You can also allow certain apps to show *detailed* or *quick* status while in the lock screen. Up to *1* app can show detailed status while up to *7* apps can show quick status. Finally, you can adjust the screen timeout and screen saver settings in this section as well.

Themes

Here, you can save or load themes or your entire current personalization preferences by going to "Theme settings". Aside from that, you have the following options under this section:

- **Advanced sound settings** – Clicking this will automatically bring you to the *Sounds* tab of the system-wide Sound options. You can choose from preset 'Sound Schemes' or individually select what sound to play during system events such as *calendar reminders, notifications, new mail notifications,* and so on.

- **Desktop icon settings** – This will allow you to manage which desktop icons should appear. Simply select the checkbox for the icons you want to appear. You can change their icons, restore the default icons, or select "Allow themes to change desktop icons" if you have installed themes that change desktop icons.

- **Mouse pointer settings** – This will open up the *Mouse Properties* window where you can access your pointing device's button configuration, pointer options, scroll wheel options, and hardware information.

Happy with how Windows 10 looks? Save your theme so you can quickly restore your computer to this state whenever you make further modifications! Go to 'Theme Settings' and look for your current theme. If you haven't saved this theme yet, it should be called 'Unsaved Theme' under 'My Themes'. Right-click this theme, choose *Save theme,* input a name for your new theme, and press *Save.* There is no limit for the amount of themes you can save, so feel free to create as many as you want.

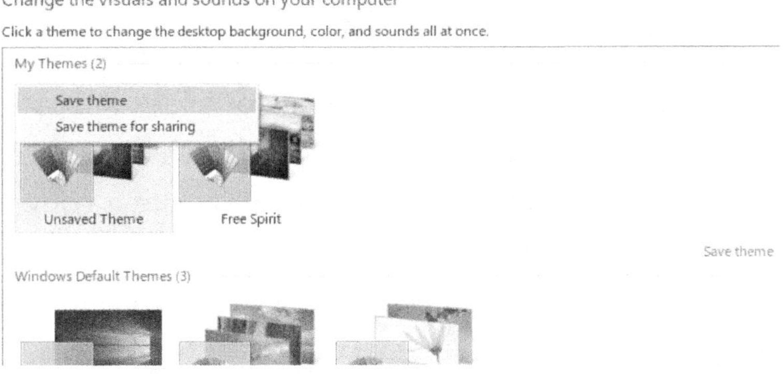

Choosing a Color for your Title Bars

As stated in *Chapter 2 – Getting Started with Windows 10,* you may notice that no matter what color scheme you choose for your theme, you'd still have white title bars for most applications except for a few (OneNote, Twitter, etc.). This isn't a bug. Rather, it is the intended design made by Microsoft to establish a unified and simplistic look for all users.

However, you can still get colored title bars to have even more personalization for your Windows 10 device. Note that the following fix may require the help of someone with reliable knowledge on how operating system files work. If you are already content with the way Windows 10 looks after all the personalization steps you've done, feel free to skip this part.

Here are the steps you need to follow if you want to use colored title bars for your Windows 10 device:

1. Using File Explorer, go to C: (Local Disk) → Windows → Resources → Themes. A quick way to get here would be to open any folder and insert this in the address bar: C:\ Windows\Resources\Themes.

2. Create a copy (Ctrl + C) of the "aero" folder and paste (Ctrl + V) it on the same folder. You will be informed that you need administrator permission to create a copy of this folder here. Simply click on 'Continue'. When prompted that you need administrator privileges to copy a certain

file, select "Do this for all current items" and click 'Skip'.

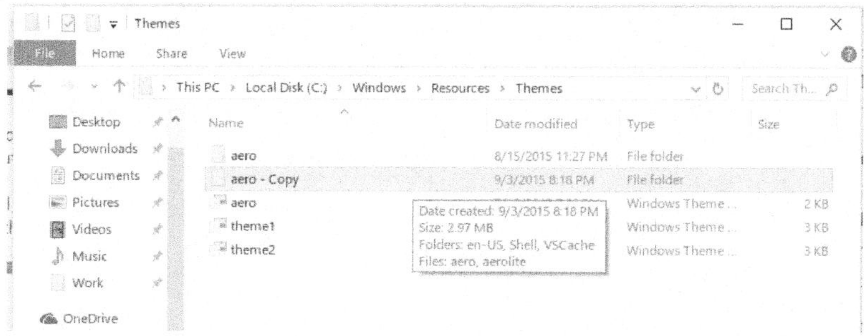

3. Rename the new folder with anything you want. Something like "aerocolor" would be appropriate.

4. Open this new folder and change the filename of 'aero.msstyles' to 'color.msstyles'. When you are presented with another *User Account Control* (UAC) message, simply click 'Continue' to proceed. Remember that if file extensions are not displayed in your folder options, this file (along with all the other files in this guide) will only show as the filename. In this case, it'll be just 'aero'.

5. Open the 'en-US' folder and rename 'aero.msstyles.mui' to 'color.msstyles.mui'. Keep in mind that 'en-US' is the language code for English users, so the folder could be named differently if you're using another language.

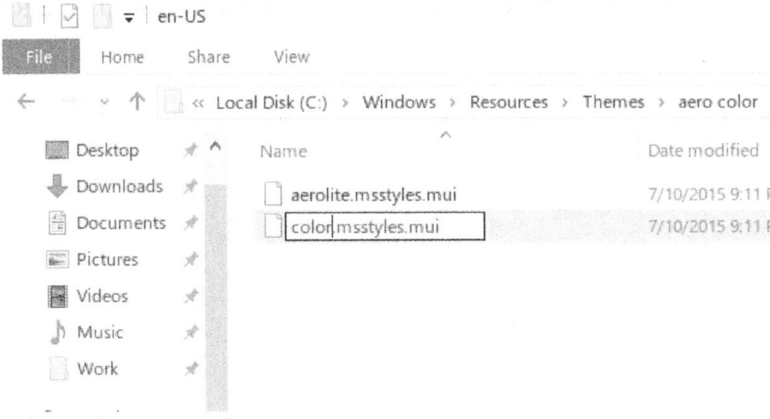

6. Return to the main 'Themes' folder and create a copy of 'aero.theme'. Paste it in another location and rename it to 'color.theme'.

7. Next, right-click on 'color.theme' and choose "Open with…". Click on 'More apps', look for 'Notepad', and then click OK. **DO NOT** select "Always use this app to open .theme files"!

8. Doing the step above should open a notepad file. Under the [VisualStyles] block (4th from the last), change the value for 'Path' from %ResourceDir%\Themes\Aero\ Aero.msstyles to %ResourceDir%\Themes\aerocolor\ color.msstyles as shown below:

```
[VisualStyles]
Path=%ResourceDir%\Themes\aerocolor\color.msstyles
ColorStyle=NormalColor
Size=NormalSize
AutoColorization=0
ColorizationColor=0XC40078D7
```

9. When done, save and close notepad. Cut the modified 'color.theme' file and paste it back to the themes (C:\ Windows\Resources\Themes\) folder. Remember to just press 'Continue' with every UAC message.

10. Double-click color.theme to activate it or choose it in 'Theme settings' (named Windows). Doing the steps above should finally allow you to have colored title bars. Remember that in case you're unhappy with the results, you can easily switch back to the intended white-title-bar design by picking another theme.

Bonus: Creating a Custom Dock

You can use a free, third-party *dock* application which will improve the navigation experience from your desktop by creating an organized set of icons. This is regarded as one advantage of *Mac* operating systems over Windows, since Windows never really implemented a built-in feature for docks.

Here is a list of the best dock applications for Windows 10, which you can download:

- **RocketDock** – http://rocketdock.com

- **XWindows Dock** – http://xwdock.aqua-soft.org

- **ObjectDock** – www.stardock.com/products/ObjectDock

CHAPTER 5 – DOWNLOADING AND SETTING DEFAULT APPS

In this Chapter:

***Introduction to the All-New Windows Store*

***Setting Default Apps for Windows 10*

***Adding Optional Features and Uninstalling Apps*

The Windows Store is another major new feature of Windows 10. Here, you can download universal apps that are accessible to all modern Microsoft devices. You can get started with the Windows Store by pressing the Store icon from the Start menu or taskbar, just look for the app with the shopping bag icon (⊞).

Bear in mind that Windows Store will use your Microsoft account, which is essential for purchasing apps. This is why it is important for you to set up your payment options in your Microsoft account.

In Windows Store, you can immediately start browsing for software you need. Take note that these are categorized into *apps, games, music,* and *movies.* Simply click any of these categories or use the search bar on the top-right corner to look for specific apps.

You should be able to see whether or not an app is paid or free as you browse. Some apps or games have *free trial* versions, which have limited functionality or accessibility. Just click on the appropriate button below the description to start downloading something from the Windows Store:

Removing an App from a PC

Remember that you can login to your Microsoft account at any time to view your transaction history. Once you purchase something from Windows Store, it will automatically be tied to your account. You can download the same paid app, game, music, or movie into your other devices as long as you use the same Microsoft account. Just bear in mind that some apps have a maximum limit of allowed PCs.

In case you want to use an app on another PC but already reached the maximum limit, you can remove that PC from your Microsoft account, which will also free the app license for that PC. To do this, go to your Microsoft account by going to the Windows Store, clicking on your account picture next to the search bar → Settings → "Manage your devices" under 'Account'.

This will bring you to a page which will list all the Windows devices you own with apps and games from the Windows Store. Find the PC you want to remove from your account and click 'Remove'.

Disabling Automatic Updates for Apps

You can disable automatic app updates by going to the same settings page where you can manage the devices linked to your account. Simply click the switch under "Update apps automatically" and set it to off. This can help save internet usage especially if you're using a connection with added data charges or if you have a slow connection and often need more bandwidth for something more important.

Setting Default Apps for Windows 10

You can set default apps when accessing certain functions of your PC. Most of these functions already have default apps set by Microsoft, but you are free to replace them with apps you've downloaded from the Windows Store or from other third-party sources.

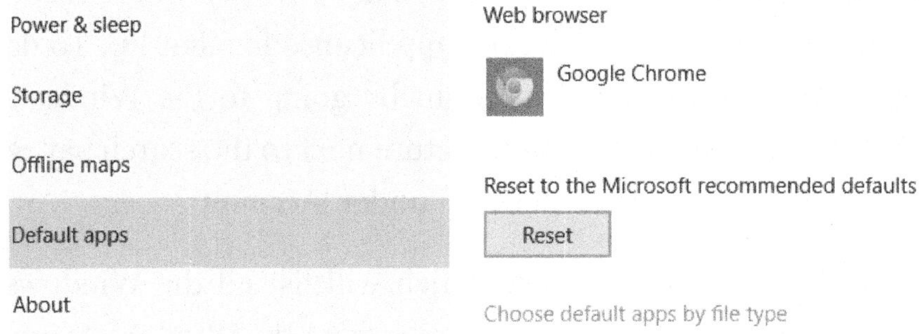

To get started with setting default apps, go to Start → Settings → System (Display, notifications, apps, power) → Default Apps. Here, you can set default apps for your *calendar, email, maps, music player, photo viewer, video player,* and *web browser*. In case you want to restore the default apps to the ones recommended by Microsoft, simply click on the 'Reset' button under "Reset to the Microsoft recommended defaults".

Adding Optional Features

You can manage the optional features installed in your Windows 10 device by going to Start → Settings → System (Display, notifications, apps, power) → Apps & Features →

"Manage optional features". Often, the preinstalled features with your operating system are essential for major features, such as Cortana, handwriting, and so on. This is why altering or removing these preinstalled features is not recommended.

On the other hand, you can go ahead and install additional features by clicking "Add a feature". This will take you to a list of all available optional features in Windows 10, which are primarily input and speech features for other languages. This is useful for multilingual users.

Uninstalling Apps

The quickest way to uninstall an app is to open the Start menu, go to All apps, right-click the app you want to remove, and then choosing 'Uninstall'. Select uninstall again when the verification message appears.

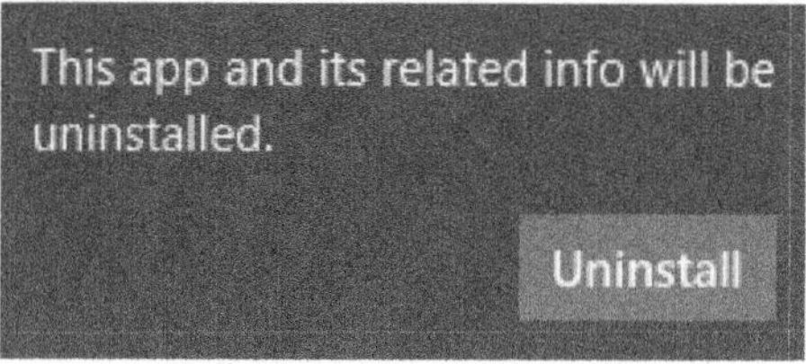

Alternatively, you can go to Start → Settings → System (Display, notifications, apps, power) → Apps & features, clicking on the app you want to remove, and then choosing 'Uninstall'. Keep in mind that you may also uninstall third-party programs this way.

CHAPTER 6 – "HEY CORTANA"

In this Chapter:

***Setting up your Digital Assistant – Cortana*

***Making Cortana respond to Voice Commands*

***Fun Tips and Tricks with Cortana*

 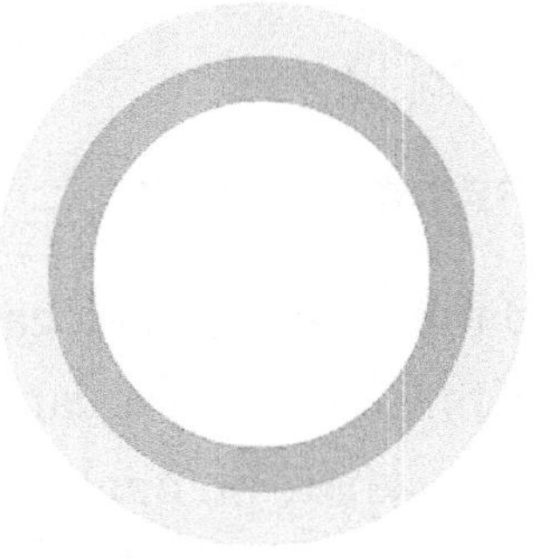

There are many new features of Windows 10 that is anticipated by many. Out of all these brilliant additions, Cortana takes the cake as being the single factor that made the Windows 10 experience stand out from previous operating systems. Although 'she' was available with a beta version in Windows 8.1 devices, it was only with the release of Windows 10 when Cortana was implemented as a core feature.

Cortana – taken from the AI character in the *Halo* game franchise – serves as an *intelligent personal assistant,* not unlike her role in the game. Of course, there were already personal assistants in other platforms before Cortana's release such as *Siri* from Apple and Google Now. Being the newest personal assistant in the market, Cortana definitely brings a lot of new things to the table. And one of the most notable capabilities of Cortana is the ability to decipher spoken words even with the presence of background noise. She can also listen to these words with impressive accuracy and speed – at least, a lot better than other personal assistants. And when all else fails, the least Cortana can do is to use Bing search for questions or commands she doesn't know.

Setting up Cortana

By default, Cortana is located in the *taskbar* right beside the Start menu button. She can be hidden or shown by right-clicking on the taskbar, highlighting 'Cortana', and choosing between 'Hidden' and "Show Cortana icon". You may also hide or show Cortana's search box from the same menu. However, keep in mind that if "Use small taskbar buttons" is enabled in "Taskbar and Start Menu Properties", then the search box will always be hidden. You can access this option by right-clicking an empty spot in the taskbar and clicking 'Properties'.

Starting Cortana for the first time will prompt her to ask for your name. You can use any name you want, but it is more appropriate for you to use your own name or nickname since Cortana will be addressing you with it. Of course, you can change your name anytime you want. This step will be discussed shortly.

Making Cortana Respond to "Hey Cortana"

Cortana has a built-in feature which allows her to respond to the user when saying "Hey Cortana". To enable this feature, access Cortana's settings by left-clicking Cortana, clicking the 'Notebook' icon from the menu, and then selecting 'Settings'.

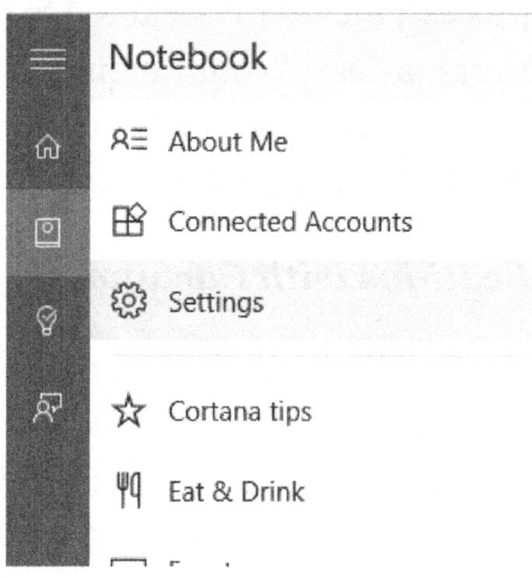

Look for the option; "Let Cortana respond to 'Hey Cortana'" and enable it. However, remember that this will consume more battery since Cortana will begin actively listening for the phrase "Hey Cortana". It is wise to turn off this feature from time to time, especially if you're running your Windows device with a battery.

If you want Cortana to respond to a voice command without enabling the "Hey Cortana" feature, then you can simply click on the *microphone* icon (🎤) next to "Ask me anything", which serves as the *search* bar for your Windows 10 system.

Making Cortana Respond to Only You

If you want more personalization, you can setup Cortana to respond only to your voice by going to Cortana → Notebook → Settings and by clicking "Learn my voice" under "Respond best to". In this process, the user is required to read *six* phrases, which will help Cortana 'learn' your unique voice. Simply press 'Start' to begin the learning process and read the phrases as clearly as possible when Cortana tells you to do so.

Running Applications with Cortana

The easiest way to test Cortana's voice recognition abilities is to use your voice to run applications. However, this is limited only to *certain* applications. For example, let us try running *Microsoft Edge* through Cortana. To do this, activate Cortana by saying "Hey Cortana". Wait for the 'beep' or check if Cortana's search bar says "Listening". Next, say "Run Microsoft Edge/ Open Microsoft Edge" in a natural and audible manner. Cortana should confirm the command by telling you that she's opening Microsoft Edge.

You can also turn some of your PC's features on or off through Cortana. Simply use the command "Turn on <feature>" to do this. For example, saying "Hey Cortana" and then "Turn on Bluetooth" will automatically turn on Bluetooth connectivity in your device.

Setting your Favorite Places

To make Cortana assist you better, you should specify your *favorite places* and *interests*. First, you can set favorite places by opening Cortana → Notebook → About me → Edit favorites under 'Favorite places'. Here, you can set specific places such as your school, work place, favorite restaurant, your parents' house, and so on. This will help Cortana assist you in reaching those places by providing traffic reports and directions.

Specifying your Interests

With the help of *Bing,* Cortana can bring you news and articles you care about right on her home page. To do this, you need to turn on the tracking of a broad category of your interest. For example, you can go to Cortana's notebook, choose "Eat & drink", and specify your dining preferences so that Cortana can provide you with recommendations in the future. Keep in mind that Cortana also learns your preferences automatically based on the things you search for, but this is a faster way for her to pick up on the things you're interested in.

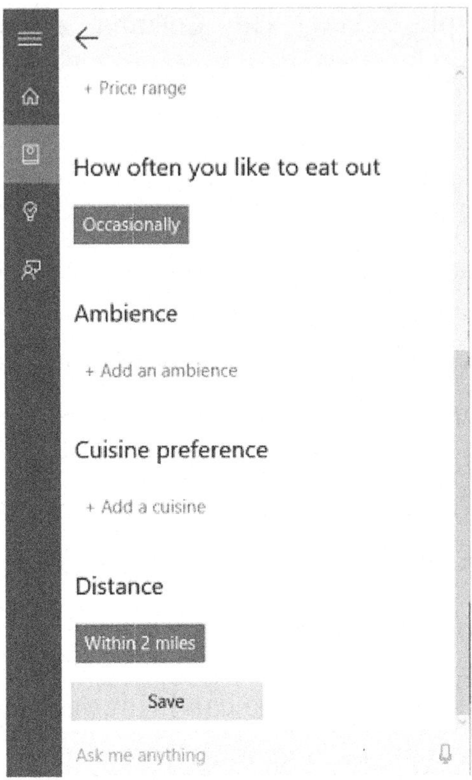

There are special options you can do with each specific interest. In relation to the example above, you can specify how often you eat out, the type of ambience, cuisine, price range, and distance you prefer when dining. There are plenty of other things you can personalize in Cortana's notebook. For example, you can track the movement of stocks you are invested in thru setting 'Finance' preferences – or stay updated with the activities of your favorite sports team thru 'Sports'.

To walk you through the entire process, suppose you want Cortana to track the NBA team, the *Golden State Warriors*. First, go to Cortana's Notebook and click on 'Sports'. Under "Teams you're tracking", click on 'Add a team' and type in; 'Golden State Warriors'. On the list that appears, click on 'Golden State Warriors, NBA'. The next page should look like:

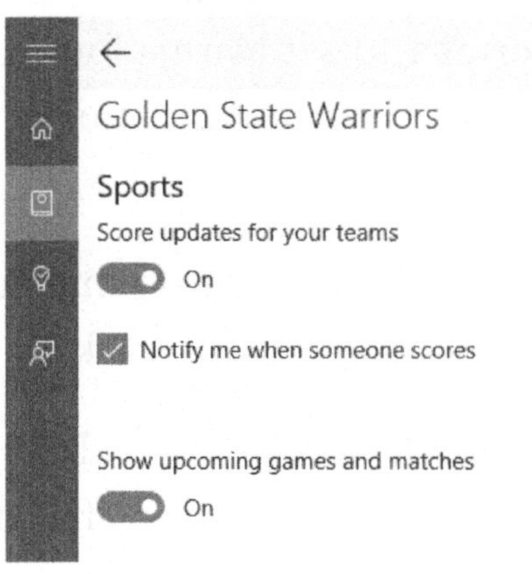

Here, you can specify the updates you want to receive regarding the Warriors. You can have Cortana inform you whenever a player scores, or show information on any upcoming games. You can track as many sports team as you want with Cortana.

Scheduling Events, Alarms, and Reminders with Cortana

You can tell Cortana to schedule reminders for you through the clock or calendar app. To do this, open Cortana and go to 'Reminders'; represented by the light bulb icon. Look for the plus (+) sign and click it to schedule a new reminder. Simply fill in a name for the reminder along with the people, place, or time associated to it.

You can also use a voice command to have Cortana schedule reminders for you. There are a few ways to do this. For example, you can say "Schedule meeting tomorrow at 2PM" or "Remind me to meet Robert tomorrow at noon" to set different reminders. You can also tell Cortana to set alarms for you. For example, you can tell Cortana; "Wake me up in 30 minutes" or "Schedule break in 15 minutes".

If you want to view your schedule or the reminders you've set so far, you can say "What am I doing this week?"

There are plenty of other voice commands you can use with Cortana. Here are some of the most useful ones you should remember:

- Navigate to "Place"

- Cancel "Event/Reminder"

- When is my next appointment?

- Set alarm for "Time"

- Take me home/How do I get home

- Show me "restaurant type" nearby

- Where is the nearest gas station?

- Show me directions to "Place"

- How's the weather today?

- How's the weather in "Place"

- Play "song name/artist/genre/playlist/album"

- What song is playing?

- Show me "today's news/the local news/the headlines"

Integrating Third-Party Apps with Cortana?

In addition to all these features, you can also integrate *third-party apps* with Cortana. This also applies for *universal apps* available through the Windows Store. And while it is originally more functional in Windows phones, Cortana's third-party integration with third-party apps for Windows 10 continues to improve.

Having Fun with Cortana

Another notable trait of Cortana is her natural-sounding voice, which works incredibly well when you ask general questions like "Who is the president of the United States". And in addition to lots of functionality, she is also capable of some *humor*. First, you can say "Tell me a joke", and she will abide by telling you a random joke from her database.

To access some of Cortana's humorous features, you can try the following commands:

- Who is your father/Who's your daddy?

- Who is your mother?

- Sing me a song

- What does the fox say?

- What is love?

- What are you wearing?

- What do you look like?

- Where do you live?

CHAPTER 7 – PREPARING AND MANAGING YOUR DEVICES

In this Chapter:

***Managing your Devices*

***Pairing Bluetooth Devices and Transferring Files*

***Storage Locations and AutoPlay*

Now that you have your personalized your Windows 10 device as well as set up your personal assistant, it's time to go back to more settings which will help you do more with your new operating system. First, you should learn how to add and manage the devices you connect to your PC.

Adding New Devices

You can manage all your devices such as printers, USB input devices, monitors, and so on by going to Start → Settings → Devices (Bluetooth, printers, mouse). Normally, devices already connected to your Windows 10 device will automatically appear here in the appropriate section.

If you recently connected a new device, promptly click 'Add a device' or 'Add a printer or scanner' – depending on the device you've connected – and wait for Windows to search for the necessary drivers. Keep in mind that drivers are required to make sure your device will work properly. Updating to the newest drivers will fix most device issues, so be sure your drivers are updated through here or through *Windows Update.*

Enabling Updates over Metered Connections

Whenever possible, Windows 10 will automatically download the newest drivers for new devices. This is intended to allow the user to focus more on productivity while the operating system works in the background to ensure full functionality. But if you have access to metered connections that incur extra charges with data usage, you should consider leaving "Download over metered connections" off. You should only turn this *on* in situations where you really need to use a new device and a metered connection is the only one available.

Pairing Devices with Bluetooth

If your computer comes with Bluetooth connectivity, then you should first modify your settings as well as *pair* with your devices before wireless functions such as file transfer can be done. First of all, make sure Bluetooth is turned on. You can do this via the Action Center or by telling Cortana to "Turn Bluetooth on".

Going back to Start → Settings → Devices (Bluetooth, printers, mouse), wait for Windows to detect your external Bluetooth device until it appears on the list of devices. It normally takes only a few seconds to detect your Bluetooth device. If not, make sure Bluetooth is *on* in your device and also check if it is discoverable.

Once your device appears on the list ("Ready to pair"), select it and click the 'Pair' button. A *passcode* should be displayed to verify the device being connected. Normally, both passcodes should match in both devices, but you should double-check the passcodes for safe measure. If the passcodes match, click 'Yes'. Also confirm the pairing in your Bluetooth device.

Sending and Receiving Files via Bluetooth

To begin transferring files via Bluetooth with a paired device, go to the taskbar and look for the Bluetooth icon.

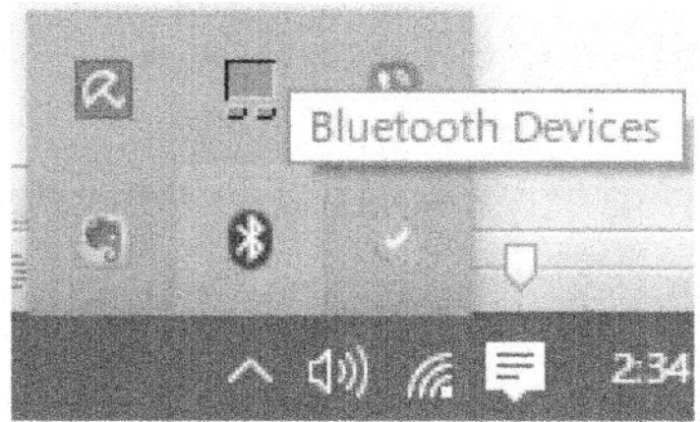

Click on this icon and choose 'Send a File' or 'Receive a File' – depending on what you need done. When receiving a file, your Windows 10 device will wait for a connection with a paired Bluetooth device. Simply send the file you want to transfer from the Bluetooth device and the rest of the process will complete automatically.

When sending a file, clicking on 'Send a File', select the device you're trying to send to from the list, and click 'Next'. Click on 'Browse' and look for the file or *files* you are trying to send. Keep in mind that you can select multiple items by holding down 'Ctrl' as you click. Once the files are selected, click on 'Next' to proceed with the transfer process. Be sure to accept the file from the Bluetooth device you're trying to send to. Click on 'Finish' after successfully transferring your file.

Other Bluetooth Settings

You can access other Bluetooth settings by clicking 'More Bluetooth options' under related settings. Here, you can change the discoverability of your Windows 10 device to other Bluetooth devices, the notification settings, and the visibility of the Bluetooth icon in your taskbar notification area.

Managing your Storage Locations

You can access your storage settings by going to Start → Settings → System (Display, notifications, apps, power) → Storage. Here, you can view all the currently connected storage devices. You may also click on a storage device to see the current usage allocation such as for apps, games, music, videos, mail, and so on.

Save locations

Change where your apps, documents, music, pictures and videos are saved by default.

New apps will save to:

This PC (C:)

New documents will save to:

This PC (C:)

New music will save to:

This PC (C:)

You can also specify the default storage locations for specific file types. These can be set under 'Save locations'. Simply click on the dropdown menus for each file type and choose the storage device you want. This is useful if you have backup volumes and want to manage your data more securely.

Changing AutoPlay Behavior

Whenever connecting new media devices, Windows performs a specific action depending on the user's choice – as with previous Windows versions. You can set the most appropriate action whenever you connect a new device, but you can set these defaults manually through Start → Settings → Devices (Bluetooth, printers, mouse) → AutoPlay. Simply choose from the dropdown menu under 'Removable drive', 'Memory card', and other detected devices in your system.

CHAPTER 8 – INCREASING PRODUCTIVITY WITH WINDOWS 10

In this Chapter:

*** Using Windows 10's Refined Snap Feature*

***Using Virtual Desktops*

***Power and Performance Management*

Windows 10 has many features that can increase productivity. After making sure your devices are up and running, let's get started with getting things done with your new operating system. In previous Windows versions, you can 'snap' Windows to the side of your screen so they will occupy half of it – automatically adjusting to fill this space. In Windows 10, this feature is refined by allowing you to *choose* which window to snap on the other half of the screen.

In case these features aren't working, make sure this feature is enabled in Start → Settings → System (Display, notifications, apps, power) → Multitasking. Simply turn on *all* the features under 'Snap'. Of course, you may choose to turn off this feature if you prefer to manually snap each window every single time.

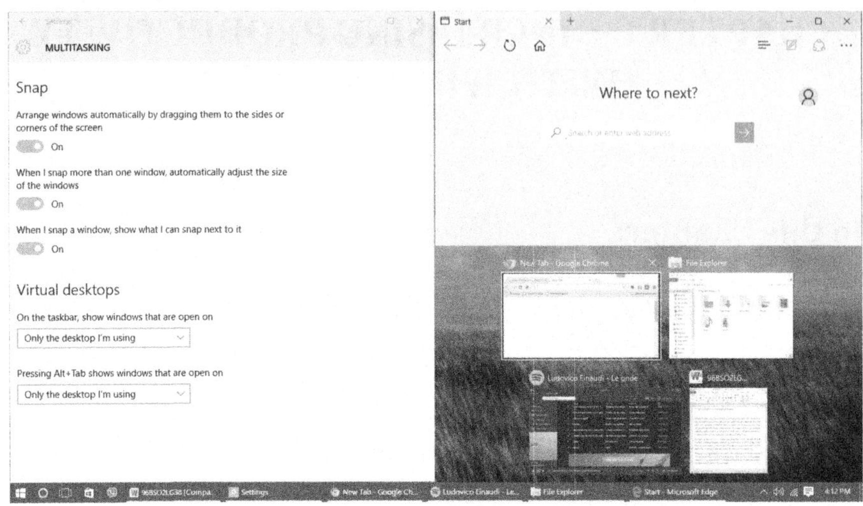

You may also snap windows to the corner of the screen for it to occupy a quarter of it. And if the third feature is turned on under 'Snap' ("When I snap a window, show what I can snap next to it"), Windows will also show you what you can snap on top or at the bottom of this window.

Using Task View

There are multiple ways to switch active windows in Windows 10. Aside from pressing Alt + Tab to choose between open windows and applications, you may also use the *task view,* which can be accessed by clicking the 'Task View' button on the taskbar (▯) or by pressing the Windows key + tab. Task view displays a list of everything running such as windows, applications, or games. Selecting something from task view brings it on top.

Using Virtual Desktops

You can multiply your workspaces in Windows 10 using *virtual desktops*. Virtual desktops or just desktops can be added by going to *task view* and clicking 'Add desktop' at the lower-right corner of the screen. Take note that this will automatically designate your current desktop as 'Desktop 1' while creating an empty desktop called 'Desktop 2'.

For example, you can use your main desktop (desktop 1) as your workspace and create a virtual desktop for your social media apps. In other words, this will help you quickly organize your workload or shift between productivity and rest.

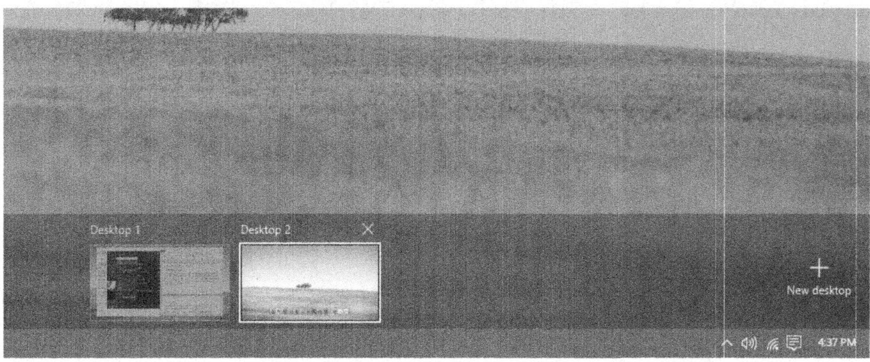

Of course, you can create more virtual desktops depending on how many you need. To close a desktop, click on the 'X' button on top of it. Remember that this will *move* anything you have opened in that desktop to the next available desktop.

You can also move apps and windows into other desktops by opening the tab view, clicking and holding on the app or window you want to transfer (from the list), and dragging it to another virtual desktop. This is useful for reorganizing the apps you have on each desktop. You can also drag something into the 'New desktop' button so you can create a new virtual desktop with that app or window already moved in it.

Accessing Apps and Windows in All Desktops

If you have multiple windows or apps running across different desktops, you can choose to show all of them in the taskbar. You can enable this in Multitasking (Start → Settings → System (Display, notifications, apps, power) → Multitasking) under 'Virtual desktops'. Simply select 'All desktops' in the dropdown menu under "On the taskbar, show windows that are open on". You may also choose 'All desktops' under "Pressing Alt + Tab shows windows that are open on", which will show everything across all desktops when you press Alt + Tab.

Other Tips for Increasing Productivity with Windows 10

If you have been using previous versions of Windows, then you can easily feel comfortable with everything Windows 10 brings into the table. Aside from the features discussed above, there are also plenty of other things you can do with Windows 10 to work faster and be more productive:

Quickly Minimizing All Other Windows

When working with Windows 10 (or any operating system for that matter), it is normal to have several windows open for multitasking. However, this can lead to the stressful sight of multiple windows open in the background as you work. Even if you're doing something in full-screen, it just feels *wrong* to know that there are plenty of other things running as you work.

One way to quickly minimize windows is to click the 'Show desktop' button located at the other end of your taskbar (opposite the Start menu button), then restoring the window you want to work with. Alternatively, you can *click and hold on the title bar of the window/app you need open, and then shaking slightly.* This will automatically minimize everything else. You can repeat the process to undo the minimization of other apps. Take note that this feature is already included in *Windows 7,* although it is still not well-known even with the release of Windows 10. There are also plenty of other tricks you can do in Windows 10 to improve your overall experience. These will be discussed in the next chapter.

Power Management with Windows 10

The power management capabilities in Windows 10 are not too different from those available in previous Windows versions. You can change when your computer will turn its display or go to *sleep mode* in Start → Settings → System (Display, notifications, apps, power) → Power & sleep. These settings will help save battery power when the device is not in use.

However, there are certain occasions when power consumption is still an issue for devices that are plugged in, especially for machines with faulty power supply units. Although unlikely, some *laptops* that are plugged in may also fail to turn off its screen even though the lid is closed. And if automatic sleep mode isn't enabled when plugged in, this can result to overheating. With this being said, it is still generally a good practice to specify a time when these features will be triggered.

Other Options for Saving Battery

Aside from power management controls, you can also help conserve battery power by using the 'Battery saver' feature in Start → Settings → System (Display, notifications, apps, power) → Battery saver. Battery saver works by restricting background activity, including background updates, processes, and notifications.

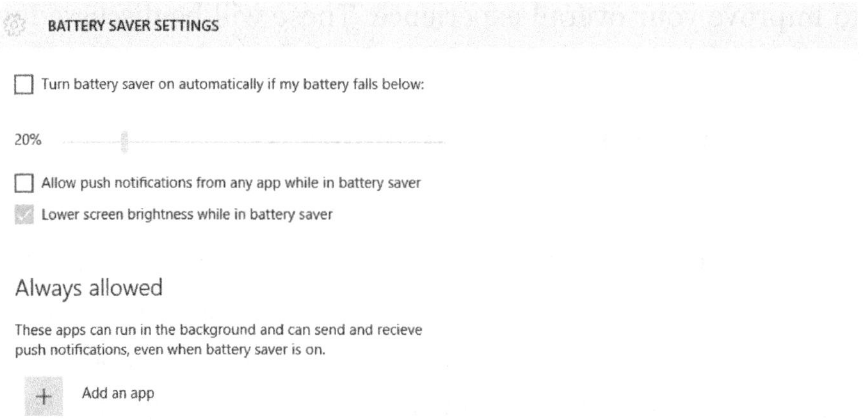

In "Battery saver settings", you can specify when the battery saver feature will be triggered according to your battery's level of power. You may also choose to allow push notifications or maintain the screen's level of brightness by selecting or deselecting the appropriate option.

Lastly, you can specify apps that will be unaffected by the battery saver feature. In other words, these are the exceptions list of battery saver. This is useful if you're trying to perform an important task with low power.

CHAPTER 9 – THE FULL WINDOWS 10 EXPERIENCE

In this Chapter:

***Recording your Gameplay Videos (and some Apps)*

***Downloading Offline Maps*

***New Keyboard Shortcuts for Windows 10*

Even though you've reached Chapter 9 of this book, you're still just scratching the surface of the things you can do with Windows 10. Although some of these features are not really essential to keeping your operating system functional, they can improve the overall experience.

Record your Gameplay

If you're a gamer, then a Windows 10 feature that can benefit you is its built-in recording software. This is available through the Xbox app's *Game bar* for Windows 10. Although intended for recording gameplay, some non-game apps may also enable the Game bar, which can be used for recording video clips.

To take photos or videos of your gameplay, simply launch the game you want and press the Windows key + G. A small panel at the bottom of the screen will show you the available options.

While the game bar is installed in all Windows 10 PCs, not all machines are capable of recording video gameplay. The hardware requirements for this feature are as follows:

1. **Intel Graphics Cards:** Intel HD graphics 4000 or newer, Intel Iris graphics 5100 or newer

2. **Nvidia Graphics Cards:** Geforce 600 graphics or newer, Geforce 800M graphics or newer, Quadro "K" series or newer

3. **AMD Graphics Cards:** Radeon HD 7000+, HD 7000M+, HD 8000+, HD 8000M+, R9+, R7+

There are also keyboard shortcuts to the main features of the Game bar:

- Record that – Windows key + Alt + G

- Screenshot – Windows key + Alt + Print Screen

- Start Recording (Video) – Windows key + Alt + R

Lastly, don't forget that you need a Microsoft account so you can access Xbox Live data. Anything you've recorded will be saved to your Xbox Live account, which can be accessed through the Xbox app.

Downloading Offline Maps

With Windows 10, you may choose to download *maps* from an online server and save map data in your device. Once saved, these *offline* maps can still be accessed even without an internet connection. You can start downloading maps by going to Start → Settings → System (Display, notifications, apps, power) → Offline maps and clicking 'Download maps'.

DOWNLOAD MAPS: ASIA

Azerbaijan	India	Kazakhstan
119 MB	Select to choose a region	419 MB
Bahrain	Indonesia	Kuwait
60.4 MB	334 MB	52.0 MB
Bangladesh	Iran	Lebanon
89.1 MB	141 MB	68.4 MB
Brunei	Iraq	Malaysia
49.3 MB	83.6 MB	208 MB
China	Israel	Maldives
Select to choose a region	231 MB	44.9 MB

The maps available for download are grouped according to *continent* (Asia, Africa, Australia, etc.). Some maps could be anywhere around 40 MB in size, while others can reach over 400. Of course, the size of the country directly affects the size of the map to download. Downloaded maps can be updated automatically or manually.

Windows 10 Keyboard Shortcut

A user's knowledge with an operating system can reliably be measured with his or her knowledge with its *keyboard shortcuts.* While the most basic functions retain their old keyboard shortcuts (cut, paste, copy, save, open, etc.), there are new shortcuts that access some of Windows 10's new features:

- **Close Virtual Desktop** – You can quickly close a virtual desktop and move any open apps to the next available desktop using Windows key + Ctrl + F4.

- **Switch Virtual Desktop** – You can easily switch between virtual desktops without accessing the Task view. To do this, simply use Windows key + Ctrl + Left or Right.

- **Snapping Windows** – Instead of dragging a whole window to the side or corner of the screen, you can now snap them with Windows key + Left/Right/Up/Down.

Other Extra Features

Here are some additional features that you need to know about Windows 10:

More Options for 'Send to' menu

You can access more options for the 'Send to' menu by holding shift before right-clicking the file or shortcut you want to send.

Pasting to the Command Prompt

When using the command prompt, you can now copy blocks of text or codes and paste them into the console. You may copy text from the console itself as well. This is useful when performing troubleshooting steps that require typing in long codes into the command prompt. Aside from the standard shortcuts (Ctrl + C, Ctrl + V), you may use Shift + Insert to paste or Ctrl + Insert to copy.

CONCLUSION

Thank you and congratulations for finishing this book!

I hope you learned lot about Microsoft's latest operating system and realized firsthand why it is called the 'Best Windows Ever'.

Remember that knowledge about anything can increase our appreciation with it. There are new things you can learn in Windows 10, which are often best learned through experience.

Again, thank you for purchasing this book and I hope you learned a lot from it!

Finally, if you enjoyed this book, please take the time to share your thoughts and post a positive review on Amazon. It'd be greatly appreciated!

Thank you and good luck!

Made in the USA
Monee, IL
08 July 2026